Disaster Diaries

Surviving the FLOOD

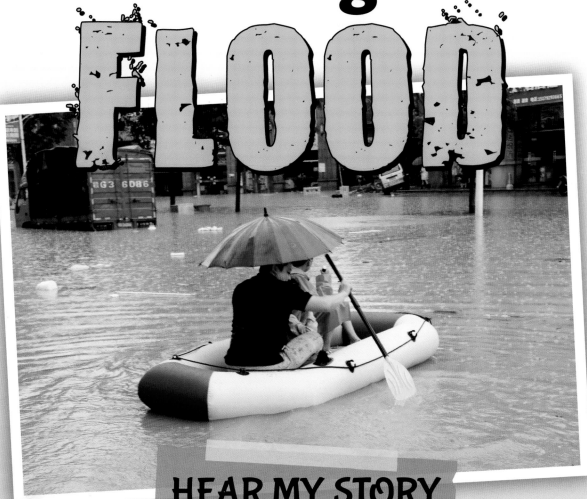

HEAR MY STORY

Heather C. Hudak

CRABTREE
PUBLISHING COMPANY
WWW.CRABTREEBOOKS.COM

Author: Heather C. Hudak

Editorial director: Kathy Middleton

Editors: Sarah Eason, Jennifer Sanderson, and Ellen Rodger

Proofreaders: Tracey Kelly, Melissa Boyce

Editorial director: Kathy Middleton

Design: Paul Myerscough

Cover design: Margaret Amy Salter

Photo research: Rachel Blount

Production coordinator and Prepress technician: Tammy McGarr

Print coordinator: Katherine Berti

Consultant: John Farndon

Produced for Crabtree Publishing Company by Calcium

Photo Credits:
t=Top, c=Center, b=Bottom, l= Left, r=Right

Inside: Shutterstock: Sk Hasan Ali: p. 18; Animaflora PicsStock: p. 14; J. Bicking: p. 10; Jose y yo Estudio: p. 27; EyeTravel: p. 19; Mikhail Gnatkovskiy: p. 8; Sean Hannon Acritelyphoto: p. 20; Brett Holmes: p. 11; Humphery: pp. 1, 12–13, 13r, 17t, 23r; Ares Jonekson: p. 4; Jianbing Lee: p. 29r; Patrizio Martorana: p. 24; Migel: p. 5; Lomtong Monrudee: p. 21; Peepy: p. 9; Pengwei: pp. 6–7; Guy J. Sagi: p. 25; Phoutthavong Souvannachak: p. 15; Wil Tilroe-Otte: p. 26; HelloRF Zcool: p. 7b; Wikimedia Commons: 坟前的丁香花: pp. 16–17, 22–23; 激动下下: pp. 28–29.

Cover: Getty Images / Stringer

Publisher's Note: The story presented in this book is a fictional account based on extensive research of real-life accounts, with the aim of reflecting the true experience of victims of natural disasters.

Library and Archives Canada Cataloguing in Publication

Title: Surviving the flood : hear my story / Heather C. Hudak.
Names: Hudak, Heather C., 1975- author.
Description: Series statement: Disaster diaries | Includes index.
Identifiers: Canadiana (print) 20200151320 |
  Canadiana (ebook) 20200151339 |
  ISBN 9780778769897 (hardcover) |
  ISBN 9780778771173 (softcover) |
  ISBN 9781427124463 (HTML)
Subjects: LCSH: Floods—China—Juvenile literature. |
  LCSH: Floods—Juvenile literature. |
  LCSH: Disaster victims—Juvenile literature.
Classification: LCC GB1399.5.C6 H83 2020 |
  DDC j551.48/90951—dc23

Library of Congress Cataloging-in-Publication Data

Names: Hudak, Heather C., 1975- author.
Title: Surviving the flood : hear my story / Heather C. Hudak.
Description: New York : Crabtree Publishing Company, 2020. |
  Series: Disaster diaries | Includes index.
Identifiers: LCCN 2019057470 (print) | LCCN 2019057471 (ebook) |
  ISBN 9780778769897 (hardcover) |
  ISBN 9780778771173 (paperback) |
  ISBN 9781427124463 (ebook)
Subjects: LCSH: Floods--China--Juvenile literature. |
  Flood damage--China--Juvenile literature. |
  Disaster victims--China--Juvenile literature.
Classification: LCC GB1399.5.C6 H84 2020 (print) |
  LCC GB1399.5.C6 (ebook) | DDC 363.34/93--dc23
LC record available at https://lccn.loc.gov/2019057470
LC ebook record available at https://lccn.loc.gov/2019057471

Crabtree Publishing Company

www.crabtreebooks.com   1-800-387-7650

Printed in the U.S.A./022020/CG20200102

Published in Canada
Crabtree Publishing
616 Welland Ave.
St. Catharines, Ontario
L2M 5V6

Published in the United States
Crabtree Publishing
PMB 59051
350 Fifth Avenue, 59th Floor
New York, New York 10118

Published in the United Kingdom
Crabtree Publishing
Maritime House
Basin Road North, Hove
BN41 1WR

Published in Australia
Crabtree Publishing
3 Charles Street
Coburg North
VIC, 3058

Contents

Floods and Their Victims

Floods happen when water overflows into an area that is normally dry. Even a few inches of water in the wrong place can have a huge impact on the land and people who live there. As a result, floods are some of the most common and deadliest **natural disasters** on Earth. No other form of natural disaster has caused more destruction across North America.

In 2018, thousands of people fled their homes after flooding in Jakarta, Indonesia.

Dangerous Force

Flowing water is a very powerful force. Floodwaters can destroy entire buildings and change the landscape. They can wash out bridges, roads, houses, and local services, such as overhead power lines. Floods often leave people homeless and cities and towns in ruins. People and animals caught in floodwaters may die.

In 2014, a storm in the Souss Massa Draa region of Morocco brought floodwaters that killed nearly 50 people, cut off more than 100 roads, and destroyed a bridge in the area.

Leveling Out

Like floods, oceans and rivers contain moving water. However, they do not have the same power of destruction as a sudden flood. This is because the **volume** and **pressure** in rivers and oceans are mostly even. In a flood, the water levels are uneven because more water builds up in some places than in others. Floodwaters rush quickly to lower-volume areas to try to level out. The water applies a lot of pressure as it moves, which creates a powerful force.

LAN'S STORY

In this book, you can find out what it is like to live through a natural disaster by reading the **fictional** story of Lan, a young girl living in China when heavy flooding took place there in 2016. Look for her story on pages 6–7, 12–13, 16–17, 22–23, and 28–29.

LAN'S STORY:
The Flood Hits

My name is Lan. I am 11 years old and I live in Xingtai in northeastern China. Xingtai is one of the oldest cities in the country. It dates back almost 3,500 years. There are many historic sites here, such as Kaiyuan Temple. The city is also an important business center. There are several big **petrochemical** and power companies in Xingtai. There are also a lot of steel and cement factories. My father works in one of them.

The factories have made Xingtai one of the most polluted cities in China. Most days, I can barely see the sky through the smog. But the area around Xingtai is very beautiful. The Taihang Mountains, Xingtai Gorges, and Xiaotianchi are amazing. My favorite is Kongshan Baiyun Cave.

The Taihang Mountains formed hundreds of millions of years ago. They stretch about 250 miles (402 km) across northern China, from north to south.

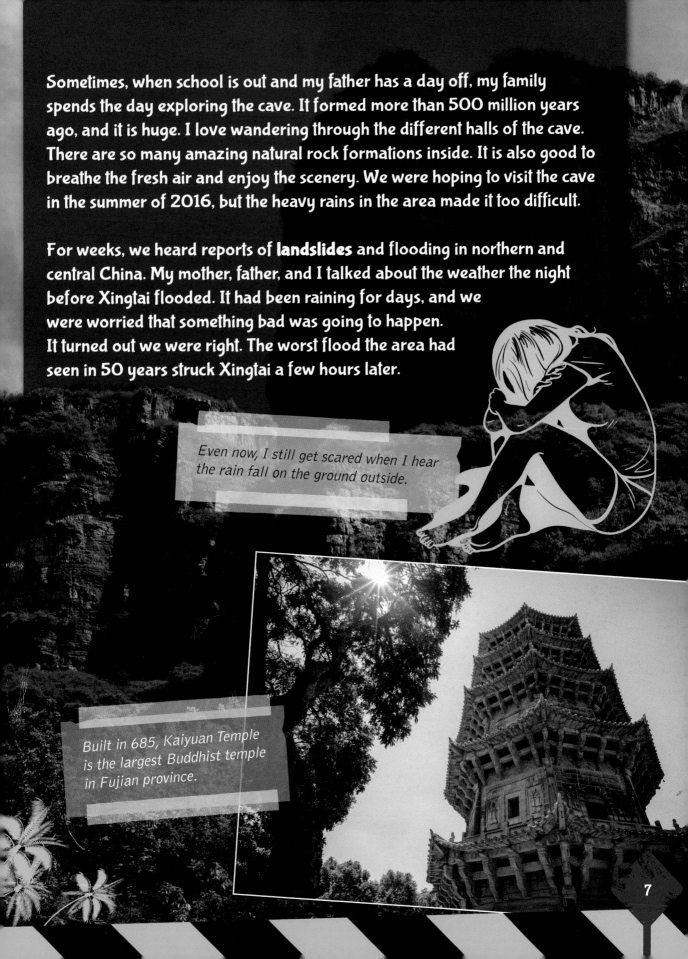

Sometimes, when school is out and my father has a day off, my family spends the day exploring the cave. It formed more than 500 million years ago, and it is huge. I love wandering through the different halls of the cave. There are so many amazing natural rock formations inside. It is also good to breathe the fresh air and enjoy the scenery. We were hoping to visit the cave in the summer of 2016, but the heavy rains in the area made it too difficult.

For weeks, we heard reports of **landslides** and flooding in northern and central China. My mother, father, and I talked about the weather the night before Xingtai flooded. It had been raining for days, and we were worried that something bad was going to happen. It turned out we were right. The worst flood the area had seen in 50 years struck Xingtai a few hours later.

Even now, I still get scared when I hear the rain fall on the ground outside.

Built in 685, Kaiyuan Temple is the largest Buddhist temple in Fujian province.

Trails of Destruction

Floods put lives at risk. On average, they kill more than 100 people each year. They also cause more than $40 billion a year in damages around the world. Sometimes, floods bring only a few inches of water to an area. Other times, the water is so deep that it can cover an entire house or other buildings. Either way, floods can have disastrous effects.

Powerless

After a flood, homes often look fine on the outside. However, there may be a great deal of damage on the inside that makes them unlivable. There may be damage to the structure that could cause the home to collapse or its ceilings to fall, for example. Standing water inside homes hides **debris** such as sharp objects, which can cause injuries.

Services, such as power, water, and phone lines, may be washed out during a flood. There may be gas leaks or live electrical wires that people cannot see behind the walls or under the water. Entire **communities** can be left without electricity and clean water for washing or drinking. People are usually **evacuated** to safety.

When floods destroy roads, it is hard to reach the communities that need help and for people to escape to safety.

Toxic Water

Floodwaters carry **contaminants** and **toxins** that pollute the land and bodies of water that were previously clean. Sewer systems often cannot handle the extra water that floods bring, so they back up into homes and city streets. This can lead to the spread of diseases. Deadly molds often grow in places that have been soaked by floodwaters.

Once floodwaters **recede**, the affected area is covered in debris, **silt**, and mud. It takes a lot of time and money to clean or replace items that are damaged or destroyed by floodwaters. Items that water can soak into, such as rugs, sofas, curtains, and mattresses, need to be thrown out because even once they are dry they could contain toxins.

Wildlife Threat

In some places, it is important to watch out for dangerous animals that have been forced out of their homes by floods. For example, alligators may be washed into yards or streets. In some places, venomous snakes sometimes seek shelter in cupboards or attics. Mosquitoes thrive in the still floodwaters in old tires, wells, and other containers.

VICTIMS OF FAMINE AND DISEASE

The 1931 Yangtze River flood is one of the worst natural disasters in history. Extreme rains led to flooding in the densely populated Yangtze River basin. More than 500,000 people living in the 500-square-mile (1,295 sq km) flood zone were forced to evacuate the area. The flood wiped out rice fields, which were the main food source for many communities in southern China. It also spread diseases through the water. About 3.7 million people died as a result, many from **famine** and disease.

Animals are at risk from flooding too. Some, such as birds and deer, can fly or run away, but others, like these sheep in Scotland, may become stranded.

How Floods Work

In many cases, it takes time for a flood to form, so people living in the flood zone have time to prepare and move to higher ground. However, some floods happen quickly and without warning, so people may not have enough time to evacuate the area.

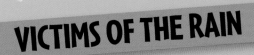

VICTIMS OF THE RAIN

The land along the Red River, which includes North Dakota, Minnesota, and Manitoba in Canada, has a history of flooding. In the 1960s, a channel 29 miles (47 km) long was built to divert water around Winnipeg, the capital of Manitoba. The area still faced major floods, but the floodway prevented tens of billions of dollars in damages. In 1997, the "flood of the century" soaked the entire Red River Valley. It led to a $627 million expansion of the floodway. Construction began in 2005 and finished in 2014.

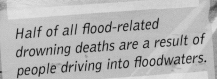

Half of all flood-related drowning deaths are a result of people driving into floodwaters.

Slow—Moving Floods

Floods caused by bodies of water, such as rivers overflowing their banks, can take hours or days before they cause major destruction. These are called slow-onset floods because they build up slowly over time. They mainly happen in flood-prone areas and can be regular occurrences. People expect them to happen and are prepared for them.

Flash Floods

Flash floods occur without warning, and people have little time to get out of the path of destruction. They happen when large amounts of water flood a normally dry area in six hours or less. The water moves quickly and can easily sweep away people and animals. It often washes out roads and bridges or blocks them with debris, making escape impossible. Flash floods kill more people than any other type of natural disaster.

Rapid Waters

Rapid-onset floods are not as urgent as flash floods, but they still happen at a fast pace. Most take place over a day or two, so people have some warning. They can take precautions and get to safety.

Hundreds of thousands of vehicles are damaged by floodwaters each year in the United States.

LAN'S STORY:
Finding Higher Ground

On the night of the flood, we went to bed at around 9 p.m. My mother gave me a kiss on the head like she always does and told me to have sweet dreams. We could never have imagined what was about to happen.

A loud sound woke me up a few hours after I had fallen asleep. I jumped out of bed to find that the floor of my bedroom was covered in water. It was rising fast. I screamed, and my parents came running. Our house was flooding and we needed to get out. My father used a chair to break the window in my room. Then we climbed onto the roof.

We sat on the roof and watched the water rushing through the city streets. The power had cut out, and the rain was coming down in sheets, so we could barely see. What we could see was horrible.

There were cars and trees being swept away by the floodwaters. Through the noise of the rushing water, we could hear people screaming for help. It was awful, and I was so scared that the floodwaters would tear our house out from under us.

Water poured into the city of Jiujiang during the 2016 flood. Many people had to travel across the city by boat.

12

We had been sitting there for awhile before emergency alarms sounded in our community. We wondered why they had not sounded sooner, because by the time they had, it was too late. We could see the water spilling out of my bedroom window. We figured it must have been almost chest deep. We could see some other families on their rooftops and hoped more people in our community made it to safety.

We spent the night on the roof, huddled together, scared and cold. When the Sun began to rise, the damage became clear. There were people clinging to trees, holding on for their lives and hoping they wouldn't be caught up in the rush of water. I wanted to help them but I knew there was nothing I could do.

As the water rose, entire streets became flooded. Many cars were stranded in the floodwater.

The Causes of Floods

Floods can happen in any place and at any time. In fact, every U.S. state and territory and every province in Canada is prone to flooding. There are many different causes of floods, depending on where they take place, the time of year, and other factors.

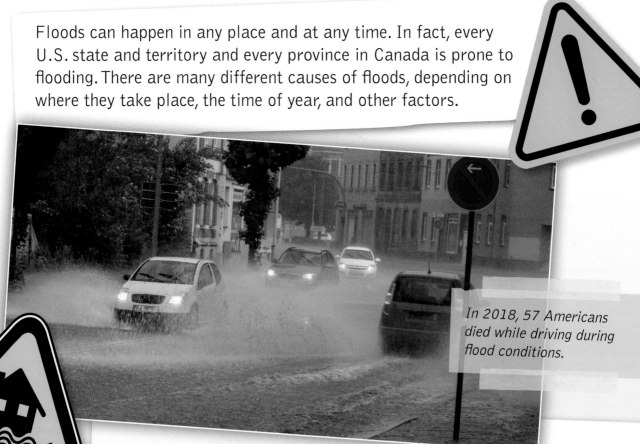

In 2018, 57 Americans died while driving during flood conditions.

Too Much Water

In some parts of the world, snow and ice build up over the winter. Usually, snow and ice melt slowly over a few weeks. Other times, a sudden change in temperature can cause them to melt too quickly, bringing too much water. Similarly, light rains that last for long periods of time or heavy rains that last a short time can bring more water to an area than normal. Too much water of any kind can cause flooding.

The **excess** water usually runs into local **drainage systems** or bodies of water, such as rivers. Sometimes, however, there is so much water that the drainage systems cannot handle it all, and they overflow. Rivers may breach, or break through, their banks, sending water into low-lying areas, or **floodplains**.

Broken Systems

Many cities and towns located in floodplains put special systems in place to try to prevent flooding. A dam is a solid barrier that is built across a river or stream. It is supposed to stop the flow of water. A levee is a structure built along the banks of a body of water to keep water from spilling over onto nearby land. Sometimes, dams and levees are not strong or big enough to hold back the water. Broken dams and levees are some of the most common causes of floods.

Storm Side Effects

Large storms, hurricanes, strong winds, or tsunamis can cause rising water levels in the ocean and other bodies of water. They can create massive waves that force seawater into coastal areas. **Climate change** is causing more extreme weather events, such as tornadoes, which bring more rain. Both of these factors can lead to flooding.

In 2018, a dam in Attapeu, Laos, collapsed. Flash floods swept through six villages, leaving nearly 7,000 people homeless.

VICTIMS OF THE DAM

Johnstown in Pennsylvania was the site of the worst flash flood in U.S. history. In May 1889, more than 2,200 people died and nearly 1,000 more were missing after the South Fork Dam unexpectedly gave way to heavy rains. On May 31, a wall of water 40 feet (12 m) high rushed through the Little Conemaugh River Valley. Within 45 minutes, it had engulfed Johnstown.

LAN'S STORY:
The Water Recedes

As soon as the sunlight was bright enough to see clearly, my mother looked me over from head to toe. I had a few scrapes and bruises on my knees from crawling on the rooftop. My mother had a gash on her leg. She had cut it on some glass from the window. My father had a few cuts but no major injuries. We knew we were very lucky. We could have died that night.

We hoped that help would arrive soon to get us off our roof, but no one came. We saw a few soldiers helping people in the streets, but most people climbed down from their roofs on their own. My father climbed down first. He went to find our ladder, but it was gone. Nothing was where it had been before. The water had carried it all away.

We struggled down from the roof and headed to the hospital because my mother needed to have her leg looked at. At the hospital, there were people everywhere. Some were very badly hurt, and we had to wait for several hours for help. A nurse finally came and cleaned my mother's wound. She stitched the gash and gave my mother medicine to stop any infection. We were told about a theater where we could take shelter.

DANGER

The 2016 flood season was the worst China had seen since 1998, with 26 provinces affected by the floods.

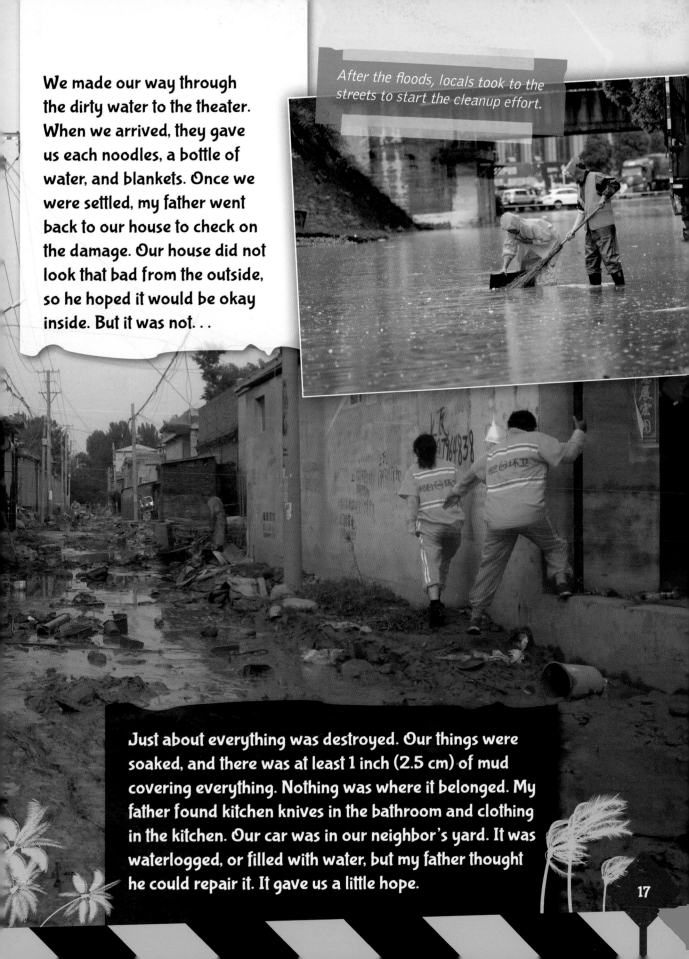

We made our way through the dirty water to the theater. When we arrived, they gave us each noodles, a bottle of water, and blankets. Once we were settled, my father went back to our house to check on the damage. Our house did not look that bad from the outside, so he hoped it would be okay inside. But it was not. . .

After the floods, locals took to the streets to start the cleanup effort.

Just about everything was destroyed. Our things were soaked, and there was at least 1 inch (2.5 cm) of mud covering everything. Nothing was where it belonged. My father found kitchen knives in the bathroom and clothing in the kitchen. Our car was in our neighbor's yard. It was waterlogged, or filled with water, but my father thought he could repair it. It gave us a little hope.

17

Where Floods Happen

Floods can happen anywhere that rain falls, which is just about every place on Earth. This is why floods are so much more common than other types of natural disasters. However, water typically moves from higher ground to lower ground, so low-lying areas near higher land or along coastlines have a greater chance of flooding.

Human–Made Disasters

Built-up areas are very likely to flood. Construction of roads, malls, parking lots, and other buildings increases the chance of a flash flood. This is because the dirt and plants that absorb and slow the flow of floodwaters have been replaced with concrete and other building materials. As a result, there is more surface **runoff**. Bodies of water that run through cities and towns are also sometimes **diverted** to storm drains. If there is too much rain, the drains can overflow and flood the surrounding area.

Much of Bangladesh is located on a floodplain just slightly above sea level, making it prone to flooding.

High-Risk Countries

Some places face a greater risk of flooding because they are located on a floodplain. Bangladesh, Vietnam, Pakistan, Indonesia, Egypt, Myanmar, Kenya, Ethiopia, Brazil, and Afghanistan are some of the countries with the highest number of people impacted by floods. Between 1995 and 2015, floods killed more than 600,000 people and left more than 4 billion injured, homeless, or in need of urgent help.

In 2019, floodwaters in the Mississippi River Delta covered more than 500,000 acres (202,343 ha) of land in six counties.

Repeat Events

There are parts of the world that are known to flood each year, and the people there rely on the floodwaters. Some, such as the areas surrounding the Tigris-Euphrates in the Middle East, the Mississippi River in the United States, and the Nile River in Egypt, are known for their farmland. Floods have been depositing nutrient-rich silt, which is good for growing crops, in these places for millions of years. In China, the Yangtze River has flooded more than 1,000 times in the past 2,000 years.

VICTIMS OF DEVASTATION

In countries such as India, Nepal, and Bangladesh, the **monsoon** season brings floods. In 2017, more than 1,200 people died and 40 million were **displaced** as a result of monsoon flooding. More than 18,000 schools were destroyed or damaged, leaving 1.8 million children with no place to take classes. The flooding was the worst in many years.

Dangerous Times

Floods can take place any time of year so long as the conditions are right. However, there are certain times of year when floods are more likely to take place. Flash floods are most common in late spring through summer. Three-quarters of flash floods happen between late April and mid-September in North America. This is because the air is warmer and more moist then. This brings heavier rainfall and more major weather events, such as hurricanes. In the Southern Hemisphere, the wet season usually falls from October to March.

Tropical storm Imelda caused flooding in Texas in 2019. High water closed a major highway in Houston, leaving cars stranded for hours.

Heavy Rains

Summer brings heavy rains and thunderstorms to certain areas. In summer, the ground is hard and dry, so the water cannot soak into it. This can lead to floods. Hurricanes are also more common during summer and fall. When a hurricane hits land, coastal areas are especially at risk of flooding from the **storm surge** that hurricanes bring.

Climate Change

Climate change also increases the chances of flooding. **Global warming** is causing more severe weather events around the world. One reason is that warmer air holds more moisture. This leads to heavier rainfall and more powerful hurricanes.

Global warming is also causing glaciers and ice sheets to melt. This has led to rising sea levels and a greater chance of storm surges. Because of these factors, scientists predict floodplains in the United States will grow by 45 percent before the end of this century as a result of global warming.

VICTIMS OF THE SEASONS

The Nile River usually floods each year between June and September. Farmers benefit from the flooding because it leaves a strip of **fertile** land on which they can plant their crops. But some years, instead of the flood, there is a drought. In 1970, the Aswan High Dam was built to control the flooding. The dam captures water during the rainy season and releases it during times of drought. However, the farmland is not as good, since the floods no longer bring nutrient-rich soil to the surface of the land.

Dams hold back floodwaters, and they provide a *reservoir* of water for farm use, drinking, electric power, and more.

LAN'S STORY:
Checking in on Family

At the theater, we found many of our friends and neighbors. We were so happy to know that they were safe, but it was not all good news. We learned that my little cousin was missing. My uncle was holding her when the rushing water swept them out of their house. He dropped her when he was struck by a piece of debris floating in the water. After the water receded, he found my aunt holding on to the side of a building. They were both cold and scared, but they searched for hours for their little girl. They came to the shelter to get food and dry clothes, and to ask others to help them search.

Many people volunteered, and we began looking through the mud and debris. We even searched the riverbank and the cornfields near their home. We met many other people who were also searching for their loved ones. Finally, we got some good news. My cousin was safe. A woman had pulled her from the water and taken her to the hospital. We were so happy to see her alive and well.

Once floodwater recedes, it leaves behind a mess of mud and damaged land.

22

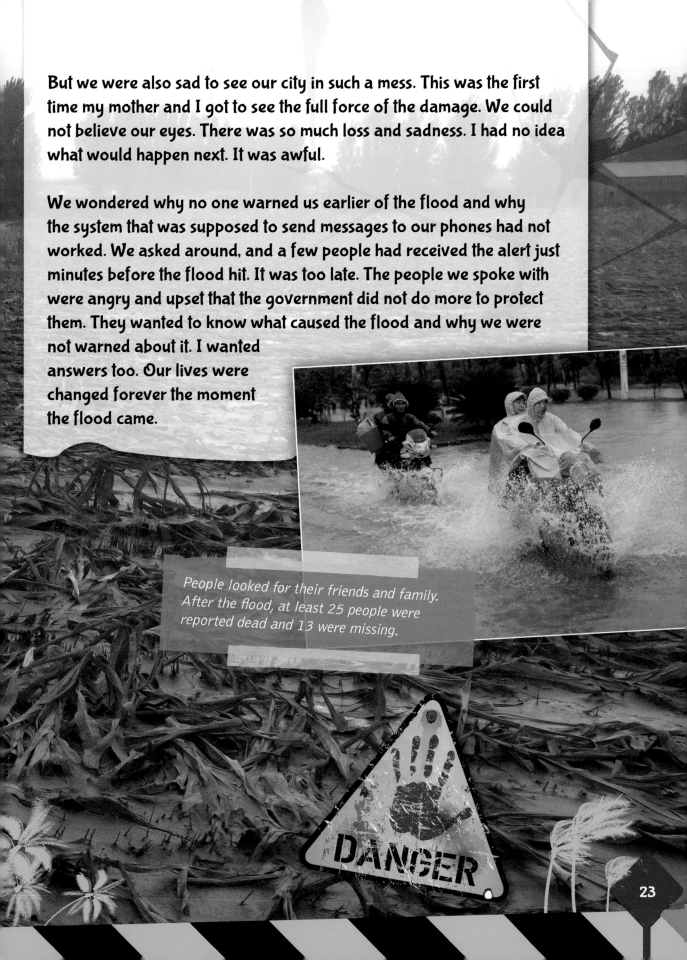

But we were also sad to see our city in such a mess. This was the first time my mother and I got to see the full force of the damage. We could not believe our eyes. There was so much loss and sadness. I had no idea what would happen next. It was awful.

We wondered why no one warned us earlier of the flood and why the system that was supposed to send messages to our phones had not worked. We asked around, and a few people had received the alert just minutes before the flood hit. It was too late. The people we spoke with were angry and upset that the government did not do more to protect them. They wanted to know what caused the flood and why we were not warned about it. I wanted answers too. Our lives were changed forever the moment the flood came.

People looked for their friends and family. After the flood, at least 25 people were reported dead and 13 were missing.

DANGER

How Science Can Fight Floods

Scientists are always looking for ways to prevent the damage, destruction, and deaths caused by floods. They use technology to predict when floods will take place and how much damage they will cause. The sooner people know that a flood may occur in their area, the more time they have to prepare.

Creating a Forecast

Hydrologists are people who study the distribution and movement of water on Earth's surface. They collect data, or information, on the amount of **precipitation**, water levels, and soil **erosion** in the area. This information helps them figure out if the water levels are high, low, or normal compared to previous years. They use the results to predict floods and help people plan for floods.

Interactive Map

The Global Flood Monitoring System (GFMS) is an online tool that monitors flood conditions around the world. It uses precipitation data from **satellites** to create a **real-time** map of flood zones. Anyone can use the computer program to see if their part of the world is at risk of flooding or how big a flood is.

Hydrologists collect water to test, and use special gauges to measure water levels in rivers, lakes, and streams.

Water is seen rising at this bridge near Fayetteville, North Carolina, after Hurricane Florence in 2018.

VICTIMS OF THE COAST

More than half of all Americans are affected by coastal flooding. Coastal and Inland Flooding Observation and Warning (CI-FLOW) technology is used to observe the effects of rainfall, river flows, waves, tides, and storm surges on ocean and river levels. CI-FLOW uses radar and rain gauge data to produce 5-minute estimates of rainfall. The estimates are used to find out the water level from storms and to predict flooding. This helps limit the loss of life and property as a result of flooding.

Airborne Lasers

In some parts of the world, scientists have attached lasers to aircraft to create maps of high-risk flood zones. The aircraft laser gathers data about an area as it flies overhead. The data is then used to predict if the area is likely to be affected by floods.

Advance Weather System

The Multi-Radar Multi-Sensor system is one of the most advanced weather forecasting products in the world. It takes data from multiple sources, such as **radar**, satellites, and rain gauges. It then combines the data from each source to create detailed maps, graphics, and charts of weather patterns. Scientists use this information to predict floods and send out warnings.

Protecting People

The best way to protect people from floods is to try to prevent them from taking place. However, it is also important to be prepared in case a flood cannot be stopped. There are many strategies communities use to help protect against flooding.

Storm surge barriers, like the one shown here, help protect low-lying areas from flooding.

Building Barriers

Some places use concrete flood relief channels to redirect floodwaters away from cities. In southern California, flood channels carry water out to sea as quickly as possible during heavy rains. In other places, barriers are used to keep water out of an area. After Hurricane Katrina, the system of floodwalls, levees, and floodgates were improved along the Mississippi River to help prevent future flooding. The floodgates can be opened or closed to redirect water.

Plant Matter

Many communities plant trees, grasses, and shrubs near shorelines. Vegetation slows the flow of floodwaters because it absorbs some of the water that would otherwise run off the ground, potentially causing a flood.

Early-Warning Systems

Many communities issue flood warnings using the news, radio, social media, and other sources to alert people to a flood. When a flood watch is sent out, it means that flooding may or may not occur. A flood warning means that a flood is about to take place or is already underway. A flash-flood warning tells people to move to higher ground immediately.

Flood Preparation

The best defense against flooding is to be prepared. People should get to know the flood zones in their community. They should also find out what causes flooding in the area so they can watch for the signs of any floods. Families should plan an evacuation route. They should make an emergency kit filled with bottled water, first-aid supplies, medications, and a battery-powered radio.

PROTECTING POTENTIAL VICTIMS

In the United States, the Global Flood Awareness System (GloFAS) is an early-warning system. It uses advanced weather forecasts and computer models to check the conditions of rivers near low-lying flood zones. The system forecasts extreme floods, so that alerts can be sent to nearby communities that are at risk. In Canada, a system is currently being developed called the Canadian Adaptive Flood Forecasting and Early Warning System (CAFFEWS).

"NO PASAR"
ALERTA
METEOROLÓGICA

Ayuntamiento de Murcia

In 2019, record-breaking rainfalls caused the Segura River to overflow in Murcia, Spain. Warning signs were posted in high-risk areas.

LAN'S STORY:
Hope after Despair

The first few days after the flood were the hardest. The streets were filled with sludge and garbage. The Sun was scorching hot. Cleanup crews worked day and night to clear the streets. There was no power, and the communication lines were down too. We relied on the shelter for food and bottled water, but there was not enough for everyone. People were suffering all around us. Most of us had no place to go.

Many people blamed the government for what happened. They said the flood was caused by human error, which had caused a reservoir to release water. But the government said that the flood was a natural disaster and that it was the result of water overflowing the banks of the Qili River.

By mid-2016, flooding had caused more than $44 billion in damages in China that year, making it one of the most expensive flooding events in history.

We may never know what really happened. The government also said that no one was injured and the area had been evacuated, but this was not true. Many people protested, asking the government to tell the truth. In total, more than 9 million people in northern China were affected by the flood. About 150 died, and 155,000 houses were damaged.

Disaster and relief workers helped us get through the early days after the flood. The government gave the city some money to rebuild, but it was not much. Insurance did not cover the damage to our home, and we had to work hard and save money to rebuild and buy all new furniture and clothes. But those are just things that can be replaced. We were very lucky that none of our loved ones died—that is what matters the most.

Many schools had to be repaired or even rebuilt after the flood.

Glossary

climate change The long-term change in Earth's weather patterns

communities Groups of people who live in one place, such as a village or a town

contaminants Substances that pollute or poison things

debris Waste or pieces of material left over from an event such as a disaster

displaced Forced to leave home

diverted Caused something to change its path or direction

drainage systems Human-made features that remove water from an area

erosion Wearing away through flowing water or wind

evacuated Cleared an area of people because it is dangerous

excess An amount of something that is more than necessary or needed

famine A great shortage of food

fertile Land that is suitable for growing crops

fictional Made up, not true

floodplains Low-lying land near a body of water that is known to flood

global warming Gradual increase in the temperature of Earth's atmosphere caused by carbon dioxide and other pollutants

landslides The fast flow of sand and rock down a slope

monsoon A wind in Asia that brings heavy rain and flooding

natural disasters Disasters caused by nature, not human-made

petrochemical Describes something that deals in petroleum or natural gas

precipitation Water that reaches Earth's surface, such as rain

pressure The force acting on a surface

radar A system used to detect radio waves to determine the angle, speed, and range of an object

real-time Something happening in the present moment

recede To move away or backward

reservoir A large lake used as a source of water supply

runoff The draining away of water from the surface of an area of land

satellites Human-made objects that send information from space to Earth

silt Very fine mud that settles at the bottom of water

storm surge The rise in the sea level as a result of a storm

toxins Poisonous substances

volume The amount of space a substance takes up or contains

30

Learning More

Learn more about floods and their dangers.

Books

Baker, John R. *The World's Worst Floods*. Capstone Press, 2016.

Hyde, Natalie. *Flood Readiness*. Crabtree Publishing, 2019.

Parker, Steve, and David West. *Natural Disasters: Violent Weather*. Crabtree Publishing, 2012.

Ventura, Marne. *Detecting Floods*. Focus Readers, 2017.

Websites

Learn more about what causes floods at:
https://kids.nationalgeographic.com/explore/science/flood/

Discover how floods work at:
https://science.howstuffworks.com/nature/natural-disasters/flood.htm

Check out the Global Flood Awareness System at:
www.globalfloods.eu

Find out what to do before and after a flood at:
www.ready.gov/floods

Index

About the Author

Heather C. Hudak has written hundreds of books for children on all kinds of topics. When she is not writing, Heather loves camping in the mountains near her home and traveling around the world. She witnessed the floods of southern Alberta, Canada, firsthand in 2013.